LIFE! SPICE IT UP!

HOW TO TRANSFORM AND HEAL WITH 5 EVERYDAY SPICES.

DR. ANDREA BLAKE-GARRETT

CONTENTS

ABOUT THE AUTHOR

Dr. Andrea Blake-Garrett believes that better health (mental, emotional, spiritual, and physical) is vital to dominating life. She decided to transform her health and wellness for life in December 2020. The result for this phenomenal woman was the shedding of **100 pounds of unwanted body fat.** She is the CEO and founder of Teamnoexcuses50, a health and wellness company that works with clients ages 15–78 to create and maintain a healthier lifestyle. Dr. Blake-Garrett is also known as The Notorious DRABG on Instagram. Her latest book titled *DOWN 100 Pounds!* is available through Amazon.com as an e-book and paper copy.

DISCLAIMER

I am not (nor ever claimed to be) a certified healthcare professional. My doctorate IS NOT in medicine or any related field. **I strongly recommend you consult your physician before starting any health and wellness journey.** *Recipes shared within this book have been used by this author for decades. The human body is complex. Individuals may react differently to certain spices.* **I am not responsible for any adverse effect you may experience if you use any recipe in this book.**

This book is dedicated to my nana Ellen (grandmother) and Papa (her son and my father), who taught me the value of spices and their uses. Nana Ellen lived to the age of 104, and Papa lived 100 years. Both lived robust and healthy lives. It is my goal to do the same.

INTRODUCTION

Growing up carefree on a beautiful island such as Jamaica, West Indies, was the most memorable experience in my life. The bright blue ocean, colorful sights, the landscape, sounds of music (reggae, gospel, Ska, steel drums) flowing through the air, and tempting smells of various foods cooking are vividly imprinted in my heart and mind. My siblings and I moved between the homes of our parents and grandparents effortlessly. I spent a lot of time in Nana Ellen's kitchen. I was her Sous Chef and resident "Pot Licker." Oh my God! The best job ever. Once the main dish (curry chicken, curry lobster, roast beef, pea soup, etc.) was placed in the serving bowls for the dinner table, all that remained in the bottom of the pot was mine to enjoy. By the time I sat at the dinner table, I was already too full to eat one bite more. There I learned how to make starch and use the remains for her amazing cassava *bami* or *bammy*. I learned how to prepare the coconut and extract oil we used for cooking, moisturizing our hair and skin, and her medicinal ointments. The remains of the coconut were used to make delicious coconut treats (*gizzarda*, grater cake, and drops). These may have different names if you are from other Caribbean islands or tropical countries.

I was my nana Ellen's helper, which suited me just fine. Her spices were my favorites. Her game for me was to take the paper or plastic bags in the kitchen and identify spices by sight, smell, and touch. Unlike many of us, she did not have labeled canisters to store her spices like those in most kitchens today. Her spices were stored on a shelf above her prep table. This way, they were easily accessible. She used spices and herbs to strengthen herself and her family as we lived on the island. She, a well-known herbalist, would treat members of our community when a doctor was not accessible or affordable. There was always a warm cup of spice tea in the morning and before bedtime. Unsurprisingly, my papa, Nana's son, was an excellent chef. They made the most delicious dishes like curry lobster, curry crab, steamed fish, vegetables, rice, and all kinds of soups. I can taste them as I write this book.

This book highlights five spices used by Nana and Papa for cooking, health, and wellness. A summary of each includes information, various uses, and a few recipes to try at home. They're most effective when included in a consistently healthy and balanced diet.

There are many books on the subject of spices on the market; thank you for choosing this one. Every effort was made to ensure it contains useful and valuable information. Please enjoy it. *Kindly submit a 5-star review on the Amazon platform.*

A LITTLE CHEMISTRY

The word chemistry may catapult you back to your school days. Do not worry! I will not go that deep. Just a few words about the body system. The Ph of your body should lean toward being more alkaline than acidic.

- **Cinnamon** (*Cinnamomum zeylanicum*) is acidic with a pH of 2–3.
- **Clove** (*Syzgium aromaticum* or *Eugenia caryophyllis*) is acidic with a pH of 3.8.
- **Curry Powder** – Since curry is a blend of spices, there is no one particular scientific name for it. pH range from 5.4–6.4.
- **Curry Leaf** (*Murraya koenigii*).
- **Ginger** (*Zingiber officinale*) is acidic with a pH of 5.6–5.9.

- **Turmeric** (*Curcuma longa L.*) is alkaline with a pH of 7.4–9.

A more acidic body can cause a decrease in the body's ability to absorb nutrients, increase weight gain, higher stress levels, and a lower immune system. Ingesting processed foods, alcohol, caffeine, and refined sugars can create a more acidic body. To create a more basic (alkaline) body system, it is suggested that fresh vegetables, fruits, nuts, and certain spices such as curry powder, cinnamon, and ginger become a regular part of the day. While neither Nana Ellen, Papa, nor their parents were chemists, they knew that a little of this and a dash of that created new flavors for spices. They were armed with oral recipes handed down through the generations, and the desire to be creative with what you have; these perfect blends of spices never disappoint any dish.

1

CINNAMON

Cinnamon
collection

There are over 245 varieties of cinnamon trees and shrubs throughout the world. What is the origin of cinnamon (*Cinnamomum zeylanicum*)? Who discovered it? Who knows? Some claim it came from Africa, while others say China. The oldest account that supports the African origin of the spice indicates that Egyptians used cinnamon in their mummification process. Therefore, I choose to believe in its African origin. Over the centuries, this unique brown spice has changed hands and locations an immeasurable number of times. It can be grown just about everywhere with a suitable climate. It is widely used worldwide in many things, including cooking, baking, perfumes, and fragrances, and the accent in wines.

Cinnamon grows as an evergreen tree. It is harvested by cutting branches and stripping the tree's inner portion of the bark. Once dried, the spice is ready to be used in stick, powder, or oil form.

This popular and versatile spice has many health benefits as well. Cinnamon's primary active ingredient is a chemical compound known as Cinnamaldehyde. Its antiseptic, analgesic, antiviral, antibacterial, and antifungal properties make it a valuable gem among the spices. Cinnamaldehyde is an essential oil responsible for the spice's flavor and scent. Several medical trials focus on cinnamon to treat respiratory infections, arthritis, diabetes, heart disease, cancer, and liver disease. It is particularly potent in the destruction of colon cancer cells. The anti-inflammatory qualities of cinnamon have tremendous value. It is effective in treating chronic inflammatory conditions such as allergic reactions like asthma and inflammation of the liver.

In helping to fight cardiovascular disease, cinnamon can reduce LDL cholesterol and help maintain the body's HDL cholesterol level. Regarding cholesterol, LDL is bad. HDL is good. For gut health, cinnamon combined with clove offers bacteria-fighting properties for treating digestive issues. Working as an enzyme blocker, it can help lower blood glucose levels.

Adding a dash of cinnamon powder to your tea, oatmeal, French toast, or porridge can undoubtedly spice up your life.

Make Your Own

Simple Cinnamon Tea

Directions:

1. Place 2 cinnamon sticks in 2 cups of water.
2. Boil for 10–15 minutes.
3. Enjoy as is with your favorite sweetener.

CLOVE

Best used whole, clove is one of the most prehistoric and valuable spices. The clove evergreen tree (*Syzgium aromaticum* or *Eugenia caryophyllis*) is grown in various exotic places such as the Comoros, China, Kenya, Sri Lanka, Zanzibar, Madagascar, Tanzania, and Indonesia. This spice is the small, nail-shaped, dried, unopened portion of the clove flower. An interesting fact is that Indonesians are not only the world's top producers of cloves but also the world's top consumers. Clove-flavored cigarettes are so popular in the country that they import cloves from Africa to supplement their crops.

The clove has an obvious culinary health value—a source of manganese, fiber, vitamin K, vitamin C, calcium, and magnesium. According to research, the clove contains eugenol, showing it is an effective local anesthetic. Historically used in oral health, some studies show that cloves have antimicrobial properties that add value when ground and mixed with regular toothpaste, promoting gum health. Others reveal that the composites in cloves may reduce cell growth and promote cancer cell death. Additional studies indicate the clove may help maintain a person's blood sugar level by increasing the body's ability to release insulin and improve the function of cells that produce insulin.

According to research, fifty-five percent of the body's daily value of manganese is reported to be in two grams of cloves. Manganese is an all-important mineral for maintaining brain function and building strong bones.

Combined with other spices, cloves make flavorful teas and dishes. The oil of clove considerably affects all kinds of food products, baked goods, meats, sauces, candies, pickles, etc. There is so much value packed in such a tiny spice.

Make Your Own

Clove-Ginger Tea

- 15–20 cloves
- 3- to 4-inch pieces of ginger (crushed, minced, sliced)
- 32 oz of water

Directions:

1. Combine and boil until only 16 oz or 2 cups of the mixture remains.
2. Enjoy warm before bed and first thing in the morning.

The benefits are many:

- Immune booster
- Detox body
- Soothe digestive tract
- Relieve stress and anxiety
- Increase antibiotic strength
- Prevent blood clots
- Increasing blood flow aid is the dilation of blood vessels

3

CURRY

SPICES

MUSTARD SEEDS

CLOVE   FENNEL   CUMIN

BASIL   PEPPERCORN   CURRY POWDER

In my nana Ellen's kitchen, there were three types of Jamaican curry powder. The colors ranged from yellow to dark orange, depending on how much turmeric she included in the blending process. How many types of curry powder are there? Well, that is up for debate. I am most familiar with Jamaican and Indian curries. Out of many comes one powerful spice. Curry is not a single spice but a combination of spices. Mix the right spices, and you, too, could create a new spice blend. Dried turmeric, coriander, cumin, black pepper, ginger, and Scotch bonnet pepper (substitute: cayenne or chili pepper) are prime ingredients in this amazingly diverse spice blend. You can curry anything: meat, shrimp, lobster, fish, vegetables, rice, absolutely anything. Add in garlic and coconut milk, and curry takes on its own unique flavor.

It is important to note that curry leaves (the herb) are not the same as curry powder (the spice blend).

The healing properties of the individual spices that are blended together to make curry give the mixture great value. Some benefits include but are not limited to:

- Anti-inflammatory
- Improving blood sugar levels
- Reduction of nausea and vomiting
- Digestive support

- Possible cancer-fighting/prevention
- Heart health (vasodilator)
- Antibacterial properties

We should all consider the power of curry powder.

Make Your Own

DIY Basic Curry Mix

- 2 teaspoons ground cinnamon
- 4 tablespoons ground ginger
- ½ teaspoon ground clove
- 1 tablespoon allspice
- 4 tablespoons ground coriander
- 4 tablespoons ground cumin
- 2 teaspoons dry mustard (yellow)
- 3 tablespoons ground turmeric
- ½ teaspoon black pepper
- 1 teaspoon ground cardamom
- 1 teaspoon Scotch bonnet pepper

**Use ground chili or cayenne pepper if Scotch bonnet is not available. Combine and enjoy!

GINGER

Ginger (*Zingiber officinale*) has been used for food and medicinal purposes for over 3,000 years. According to several reports, India is the number one producer and exporter of this spice. The spice comes from the roots that grow underground. Ginger is an independent spice. It has the greatest value when used alone or with other herbs and spices. Numerous clinical trials focus solely on ginger's benefits, attempting to better understand how this spice works. For brain function, digestion, cancer, nausea and vomiting, arthritis, and heart disease, ginger is a "go-to" spice. It is effective in patients recovering from surgery, undergoing chemotherapy, experiencing morning sickness, and experiencing PMS.

Various scientific studies show that ginger displays antidiabetic, anti-inflammatory, antioxidant, and anti-cancer properties. Natural chemicals contained within may help fend off certain prostate, pancreatic, and ovarian cancers by slowing or stopping cancer cell growth.

Ginger adds excellent value to those seeking natural alternative relief from motion sickness. Many believe it is just as effective as prescription medicine in pharmacies. Whether traveling by land, sea, or air, this super spice relieves the sick.

Ginger contains a powerful chemical compound called gingerol that has proven effective in combating infections. It successfully battles some bacterial, fungal, and viral infections most common in tropical regions.

Our brain changes as we age. Conditions such as Alzheimer's and dementia are of great concern. Studies reveal that ginger can help with overall brain function more directly and improve memory and reaction time.

Make Your Own

Simple Ginger Tea

- Fresh ginger
- Sweetener of choice

Directions:

1. A piece of fresh ginger.
2. Trim off the tough knots and dry ends.
3. Cut it into slices or dice it.
4. Put in a cup or mug with a cover.
5. Pour in boiling water and cover.
6. Wait 15 minutes and enjoy!

5

TURMERIC

Long used in ancient African and Asian cultures, this unique spice has grown in popularity. **Turmeric** (*Curcuma longa*) is a member of the same family as ginger. It is a rhizome. Traditional Chinese medicine uses turmeric to control bleeding and treat various diseases, such as asthma and respiratory inflammation. It is extensively used as an appetite stimulant and digestive in multiple sauces, a standard ingredient in curry. Herbalists praise the many benefits of turmeric preventing heart disease and certain types of cancer and managing infection and arthritis. In traditional medicine, the spice controls bleeding and treats various conditions, including asthma, skin conditions, upper respiratory tract infection, joint pain (arthritis), and digestive malfunction. The seasoning gathered from the plant's roots is now more widely known for its cosmetic uses than as a condiment or cure. This yellow spice has become extremely in demand as a primary ingredient for cosmetic products like soaps, facial cleansers, creams, etc. A popular drink of the culture in India is haldi or golden milk, a mixture of turmeric with clarified butter, pepper, and milk.

Make Your Own

Turmeric Spice Skin Mask

- 1 tablespoon plain unsweetened yogurt
- 1 tablespoon organic ground turmeric
- 1 tablespoon raw honey
- 1 lemon (juice)
- 1 glass or plastic bowl

Directions:

1. Mix ingredients into a smooth paste.
2. Apply evenly to the desired area. Mask can be used anywhere on the face, neck, and body with a makeup brush or gloves. **Caution: This will stain, so be careful.**
3. Wait 25–30 minutes.
4. Rinse off.
5. Use daily or 2–3 times per week, depending on the severity of the skin concern.

CONCLUSION

The biggest blessing of living my younger years in a multigenerational Jamaican home was the learning. The handing down of tried-and-true methods and recipes that aid in survival of the family traditions and customs —what a priceless inheritance! Take a walk to your pantry or spice rack. Chances are these super spices are sitting right there waiting to be used to add value to your health. Whatever the health challenge, with consistent use of these five (5) spices, you can begin to reclaim your natural inheritance.

EASY RECIPES

Simple Ginger Tea

- 1 thumb of fresh ginger sliced
- Sweetener of choice (honey, brown sugar, etc.)

Directions:

1. Trim off the tough knots and dry ends.
2. Wash, then cut ginger into slices or dice.
3. Put in a cup or mug with a cover.
4. Pour in boiling water and cover.
5. Wait 15–20 minutes and enjoy!

DIY Basic Curry Mix

- 2 teaspoons ground cinnamon
- 4 tablespoon ground Ginger
- ½ teaspoon ground Clove
- 1 tsp Allspice
- 4 tablespoon ground Coriander
- 4 tablespoon Ground Cumin
- 2 teaspoons dry Mustard (yellow)
- 3 tablespoon ground Turmeric
- ½ teaspoon Black Pepper
- 1 teaspoon ground Cardamom
- 1 teaspoon Scotch Bonnet pepper

Directions:

Combine and enjoy!

**Use ground Chili or Cayenne pepper if Scotch Bonnet is not available.

Sorrel – Ginger – Clove Drink

- 3 cups dried sorrel
- ½ pound fresh ginger washed and grated
- 10 whole cloves
- Brown sugar or sweetener of your choice
- 3–4 cinnamon sticks

Directions:

1. Bring to a boil for 20–25 minutes.
2. Let sit for 2–3 hours to cool down.
3. Transfer to the refrigerator—deep dark blood-red liquid.
4. Separate the juice using a strainer into a large container. If necessary, strain again until it is clear of any solids.
5. Squeeze the solid portions to collect as much liquid as possible.
6. Discard solids.
7. Sweeten and enjoy warm or cold!

Clove-Ginger Tea

- 15–20 cloves
- 3- to 4-inch pieces of ginger (crushed, minced, sliced)
- 4 cups of water

Directions:

1. Combine ingredients in a pot.
2. Boil until only 16 oz or 2 cups of the mixture remains.
3. Sweeten with brown sugar or honey to taste.
4. Enjoy warm before bed or first thing in the morning.

The benefits are many:

- Immune booster
- Detox body
- Soothe digestive tract
- Relieve stress and anxiety
- Increase antibiotic strength
- Prevent blood clots
- Increasing blood flow aid is the dilation of blood vessels

Simple Cinnamon Tea

Directions:

1. Place 2 cinnamon sticks in 2 cups of water.
2. Boil for 10–15 minutes.
3. Enjoy as is with your favorite sweetener.

Nana's Spice Tea

Cinnamon sticks added to hot tea are particularly effective.

- 2 tablespoons of fresh ginger
- 1 ounce of lemon juice (freshly squeezed)
- 2 Cinnamon sticks
- 2 Cloves of fresh garlic

Directions:

1. Set lemon juice aside.
2. Mince garlic and ginger into small pieces. Leave skin (outer layer) on the ginger.
3. Fill a pot with 4 quarts of water and boil for 15–20 minutes.
4. Place ginger, garlic, and cinnamon into the pot of boiling water and let sit for 1 hour or more.
5. ENJOY an 8-ounce glass before and after bed. Hot or cold.

Benefits:

- Kill parasites/bacteria in the stomach
- Lower cholesterol
- Lower high blood pressure

- Assist the body in maintaining optimal blood sugar levels
- Cleanse blood
- Purify organs
- Help eliminate belly fat

Pineapple & Cinnamon Tea

- 1 cinnamon stick
- 3 tablespoons of ginger or 1 thumb of fresh ginger sliced
- ¼ teaspoon of turmeric
- 1 ounce of fresh lemon juice
- 6 cups of water
- Peel of 1 medium pineapple (wash the peels to remove dirt and impurities)

Directions:

1. Boil all ingredients in water for 30 minutes.
2. Remove from heat.
3. Cover the pot and let sit 30–60 minutes longer.
4. Strain solid pieces from the mixture.
5. Squeeze as much of the liquid from the solid pieces as possible.
6. Sweeten tea to taste.
7. Enjoy a warm cup each day.

DIY Natural Medicine – Body Aches, Inflamed Joints, and More

Organic ingredients are best

- 3 tablespoons turmeric
- 3 tablespoons *ground Moringa leaves
- 3 tablespoons ground ginger
- 1/16 teaspoon (a pinch) black pepper
- 1½ tablespoons raw honey

Directions:

1. Combine all ingredients by hand in a small bowl as if kneading dough. Mix thoroughly.
2. Form in the shape of a standard capsule.
3. Take 2 capsules every 4 hours until desired relief.

What is Moringa?

In native to tropical countries like Africa, Asia, Latin America, and the Caribbean, this miracle plant is said to be an antibiotic, hypotensive, antiulcer, anti-inflammatory, and lowers cholesterol and blood sugar, among others. Moringa (*Moringa oleifera*), packed with vitamins and minerals, contains seven times more vitamin C than oranges and fifteen times more potassium than bananas. Moringa is widely used for arthritic relief, a common ailment among older people. It can be agonizing and lead to people not being able to use their joints effectively. It helps reduce the buildup of fluids and swelling within the joints. The result is noticeable arthritic pain relief. Nature is impressive as the entire tree is edible—bark, pods, leaves, nuts, seeds, tubers, roots, and flowers. An all-inclusive plant, its seed, formed into a paste, can be used to purify drinking water. Because it may cause uterine contractions, doctors recommend that **pregnant or breastfeeding women not consume Moringa.**

LIFE! SPICE IT UP! | 49

Jamaican Cornmeal Porridge

- 2½ cups of water
- 1 can of unsweetened coconut milk
- 1 cup of "fine yellow" cornmeal
- 1 teaspoon vanilla extract
- ½ teaspoon cinnamon
- ½ teaspoon nutmeg
- Salt to taste
- Allspice
- Brown sugar, condensed milk, honey, or sweetener of choice

Directions:

1. Add 2½ cups of water and 1 can of coconut milk to a 4-quart heavy-bottomed pot. Bring to a boil over medium-high heat.
2. Add 1 cup of "fine yellow" cornmeal and 1½ cups of water to a large mixing bowl and whisk until smooth. Avoid getting lumps in your cornmeal porridge. *Skip this step at your own risk.*
3. Once the liquid in the pot begins to boil gently, whisk in the cornmeal mixture, and continue to whisk for about 1 minute, ensuring there are no lumps.

4. Reduce heat to low and cover with a tight-fitting lid.
5. Cook over low heat for 15–20 minutes, stirring occasionally.
6. When there are about 5 minutes left, stir in vanilla extract, cinnamon, nutmeg, salt, and allspice.
7. Remove from heat and stir in condensed milk, sugar, honey, or sweetener.

Papa's Jamaican Curry Lobster

- 5 lobster tails, vein removed, removal of shell optional
- ½ teaspoon onion powder
- ½ teaspoon of ground ginger
- 2½ teaspoon curry powder
- ½ teaspoon garlic powder
- 1½ tablespoon cornstarch
- 1 medium onion
- 2 stalks of scallion
- 2 sprigs of thyme
- 1 medium green bell pepper
- 1 medium red bell pepper
- 1 medium yellow bell pepper
- 1 Scotch bonnet pepper (spicy, omit if necessary)
- 2–3 cloves of garlic
- 1 teaspoon of ground ginger or fresh ginger
- 3–4 tablespoons of cooking oil
- 3 tablespoons of coconut milk (omit if allergic)
- 1 cup of water

Lobster Seasoning:

- 5 lobster tails, deveined, *removal of the shell is optional
- ½ teaspoon garlic powder

- ½ teaspoon onion powder
- 2 teaspoons curry powder

Directions:

1. Wash the lobster in white vinegar or lemon juice.
2. Rinse with cold water.
3. Cut lobster tails into 4 pieces each.
4. Mix together and set aside.
5. Add salt and pepper to taste.
6. Enjoy!

Curry Sauce

- 1 green bell pepper, chopped
- 1 yellow bell pepper, chopped
- 1 small onion, chopped
- 3 cloves garlic, minced
- 3 tablespoons cooking oil
- 3 tablespoons unsweetened coconut milk
- 1–2 sprigs of thyme
- 2 stalks of scallion
- ½ tablespoon hot pepper, no seeds (optional)
- 2 tablespoons coconut milk
- 1 tablespoon Blue Mountain curry powder
- 1 tablespoon cornstarch
- 1 cup water
- Add salt and pepper to taste

Directions:

1. Heat oil in a pot on medium heat.
2. Add remaining curry powder, garlic, pepper, scallions, onions, etc.
3. Sautee 1–2 minutes.
4. Add in coconut milk, salt, and black pepper, and mix well.
5. Mix cornstarch and water. Pour mixture into pot.
6. Cook for 3–5 minutes, and enjoy!

Pepper Pot – Guyanese Style

- One 2½–inch knob of fresh ginger, peeled and grated
- ½ of a whole nutmeg, grated
- 3 pounds of pork with bone, bone-in beef, or oxtail cut into roughly 1- by 2-inch pieces **You can use a variety of meats or one type.*
- 2 teaspoons (8 g) kosher salt, divided
- 2½ teaspoons chicken bouillon, divided
- 1¼ cups cassareep, divided
- 21 sprigs of fresh thyme divided
- 24 whole cloves divided
- 3 cinnamon sticks (about 3 inches each) divided
- 6 medium cloves garlic, finely minced
- 4 fresh Scotch bonnet pepper (make as spicy as you want)
- 2 tablespoons light brown sugar
- 1 pressure cooker

Directions:

1. Season meat well with salt and ½ teaspoon of chicken bouillon. In a pressure cooker, combine meat with ½ cup cassareep, 7 sprigs of thyme, 8 cloves, 1 cinnamon stick, and 4 cups water. Bring to cook under pressure for 60 minutes.

2. Season and cook meat accordingly. Keep in mind that different meats have different cook times. Cook accordingly.
3. Stir in minced garlic, Scotch bonnet peppers, brown sugar, grated ginger, nutmeg, cassareep, orange peel, and 1 teaspoon of chicken bouillon to the pot.
4. Bring to a boil, and cook on medium-low heat for 15–20 minutes. Season with salt if needed.
5. Remove any fat from the surface of the dish.
6. Eat with bread, over white rice, or food of your choice.

Turmeric Spice Skin Mask

- 1 tablespoon plain unsweetened yogurt or
- 1 tablespoon organic ground turmeric
- 1 tablespoon raw honey
- 1 lemon (juice)
- 1 glass or plastic bowl

Directions:

1. Mix ingredients into a smooth paste.
2. Apply evenly to the desired area. Mask can be used anywhere on the face, neck, and body with a makeup brush or gloves. **Caution: This will stain, so be careful.**
3. Wait 25–30 minutes.
4. Rinse off.
5. Use daily or 2–3 times per week, depending on the severity of the skin concern.

The benefits are many:

- Lighten acne scars
- Fight signs of aging
- Moisturize dry skin
- Create a more even skin tone
- Remove dark circles around the eyes

Caution: Do not allow the mask to get into your eyes.

Turmeric Milk or Haldi

- 1 cup warm milk (almond, cashew, etc.)
- ½ teaspoon turmeric
- ¼ teaspoon pepper
- ¼ teaspoon clarified butter, avocado oil, or olive oil

Directions:

- Blend and enjoy!

The benefits are many:

- Powerful antioxidants
- Anti-inflammatory
- Manganese
- Cough/cold remedy
- Weight loss
- Reduction of joint pain

RESOURCES:

Leech, M. J. S. (2017, June 4). *10 Delicious Herbs and Spices With Powerful Health Benefits.* Healthline.

Project, E. H. (2015, October 19). *Did you know cinnamon is actually the bark of a tree? See it's fascinating farm to fork journey.* YouTube. https://www.youtube.com/watch?v=DPCsk_fN1S4&feature=youtu.be.

14 Health Benefits of Cinnamon - Facty Health. (2021, September 3). Facty. https://facty.com/food/nutrition/14-health-benefits-of-cinnamon/?style=quick.

Medicinal Spices Exhibit - UCLA Biomedical Library: History & Special Collections. (n.d.).

NCBI - WWW Error Blocked Diagnostic. (n.d.). https://pubmed.ncbi.nlm.nih.gov/27771918/.

Cinnamaldehyde - The Smell and Flavour of Cinnamon. (n.d.). http://www.chm.bris.ac.uk/motm/cinnamaldehyde/cinnh.htm.

Nag, O. S. (2017, April 25). *The World's Top Clove Producing Countries.* WorldAtlas. https://www.worldatlas.com/articles/the-world-s-top-clove-producing-countries.html.

Nag, O. S. (2017, April 25). *The World's Top Clove Producing Countries.* WorldAtlas. https://www.worldatlas.com/articles/the-world-s-top-clove-producing-countries.html.

Dr. Weil - Integrative Medicine, Healthy Lifestyles & Happiness. (2019, November 6). *Cooking With Spices: Cloves.* DrWeil.com. https://www.drweil.com/diet-nutrition/cooking-cookware/cooking-with-spices-cloves/.

Kubala, M. J. S. (2020, January 31). *9 Surprising Benefits of Curry Powder.* Healthline. https://www.healthline.com/nutrition/curry-benefits.

Curry Powder: Are There Health Benefits? (2020, December 21). WebMD. https://www.webmd.com/diet/health-benefits-curry-powder.

Dana @ Minimalist Baker. (2020, December 10). *Curry Powder Recipe |*

Minimalist Baker Recipes. Minimalist Baker. https://minimalistbaker. com/diy-curry-powder/.

Chemistry of Spices. (2008). https://catbull.com/alamut/Bibliothek/Chem- istry_of_Spices.pdf. Retrieved January 2, 2023, from https://catbull. com/alamut/Bibliothek/Chemistry_of_Spices.pdf.

Thurston, S. (2017, December 6). *Foods That Turn the Body Alkaline.* Healthy Living. https://healthyliving.azcentral.com/foods-that-turn- the-body-alkaline-12375301.html.

Sarao, C. (2011, July 8). *How to Heal Vertigo With Ginger.* LIVESTRONG.COM. https://www.livestrong.com/article/135992- heal-vertigo-ginger/.

Health Benefits of Moringa. (2019, May 23). WebMD. https://www.webmd. com/vitamins-and-supplements/health-benefits-moringa.

Moringa. (n.d.). Food and Agriculture Organization of the United Nations. https://www.fao.org/traditional-crops/moringa/en/.

BSc, A. A., PhD. (2018, May 4). *6 Science-Based Health Benefits of Moringa oleifera.* Healthline. https://www.healthline.com/nutrition/6-bene fits-of-moringa-oleifera.

Turmeric. (n.d.). NCCIH. https://www.nccih.nih.gov/health/turmeric.

Gunnars, K. B. (2021, May 7). *10 Proven Health Benefits of Turmeric and Curcumin.* Healthline. https://www.healthline.com/nutrition/top-10- evidence-based-health-benefits-of-turmeric.

TURMERIC: Overview, Uses, Side Effects, Precautions, Interactions, Dosing and Reviews. (n.d.). https://www.webmd.com/vitamins/ai/ingredient mono-662/turmeric.

www.ingramcontent.com/pod-product-compliance
Lightning Source LLC
Chambersburg PA
CBHW022106020426
42335CB00012B/849